MAKING SENSE OF METRIC

M C D Malcolm

First published in Great Britain in 1972 by
Kogan Page Limited, 120 Pentonville Road,
London N1 9JN, under the title *Introducing Metric*,
by MCD Malcolm and J. Pouvoir. Now
completely revised and updated.

British Library Cataloguing in Publication Data

Malcolm, M. C. D.
Making sense of metric. – 2nd ed.
1. Metric system
I. Title II. Malcolm M. C. D.
389'.16

ISBN 1-85091-970-4

Typeset by DP Photosetting, Aylesbury, Bucks
Printed and bound in Great Britain by
Richard Clay Ltd (The Chaucer Press), Bungay

Contents

1. Metric and SI 8
2. The Essentials 11
3. Writing It Down 19
4. Temperature 24
5. Metrication in Sport 29
6. Metrication in DIY 31
7. Metrication and the Motor Car 34
8. Metrication in the Kitchen 38
9. Clothing Sizes 42
10. International Paper Sizes 45
11. Thinking Metric 48
12. Decimals Are Not Difficult 51
13. How Much Do You Remember? 59
14. Tables 61

1. Metric and SI

Metric measurements of length, weight, etc have been in use in most countries in Europe and many other countries in the world for more than 200 years. During this time a number of local variations developed and in 1960, as a result of international agreement, a system known as *Système International d'Unités* (SI) was agreed upon and this has been, or shortly will be, adopted by virtually every country in the world for scientific purposes and for international trade.

However, a system devised for scientific purposes is not in every way suitable for everyday use and the general reader, on first encountering the International System (SI), might become very confused and might find it difficult to see the relevance of some of the terms used to normal day-to-day activity.

The units referred to in this book are those which will be encountered in normal circumstances and not necessarily those forming the basis of SI. For example, the 'official' basic scale of temperature is that named after Kelvin and is based on the absolute zero (ie the lowest temperature thought to exist) whereas for practical day-to-day purposes the Celsius scale (with water freezing at 0 °C and boiling at 100 °C) will be used and it may be commonly known as centigrade, though Celsius is technically the correct term.

Similarly, for many scientific purposes it may be convenient to measure velocity in metres per second (m/s) but for most practical purposes speed will be expressed as kilometres per hour (k/h).

The physicist might be horrified at the suggestion that mass and weight are the same thing but for those whose concern with weight is mainly either in buying or selling goods or in watching their physical health the distinction is of no importance and it will not be referred to when the units of weight (or mass) are dealt with.

To sum up: this is a straightforward book designed for the non-scientific reader and intended to convey the basic and essential information concerning the metric system as it is being adopted in the United Kingdom.

A warning

For convenience, and to make remembering easy, many of the equivalents of imperial/metric quantities shown in this book are approximate only. While these approximations will serve when used in isolation, the small inexactitudes could lead to serious errors if the rough equivalents were added together or multiplied. For example, it is safe to say that 30 miles is *about* equal to 50 kilometres but it would not be true to say that if the distance from Gatwick to Inverness is 600 miles this is equal to 1,000 kilometres. To be exact 600 miles is equal to 966 km, a difference of about 3 per cent.

The inevitability of the change

In 1965 it was decided that the standard weights and measures in use in Great Britain should be changed from the long-established but inconsistent imperial measures to the International System within a period of about ten years. That period has been more than doubled. There have been some attempts from time to time to delay or even reverse the change but the process has gone so far in a number of industries, and even in retailing, that it would be virtually impossible to halt it, especially since in the interim we have joined the European Community, and from 1992 we shall be much more closely integrated with Europe.

Nine-tenths of the world's people, including most of the Commonwealth, already use the metric system or have decided to make the change. Even in the United States, likely to be the last major manufacturing nation to make the change, the National Bureau of Standards has recommended to Congress that 'the United States should change to the metric system through a co-ordinated national program' and that 'early priority be given to educating every American schoolchild and the public at large to think in metric terms'.

For many years now children in the UK have been taught to measure and calculate in metric and today's young adults may know nothing of bushels and pecks, drachms and scruples, or even hundredweights and centals.

2. The Essentials

For everyday purposes – for shopping or house-furnishing, for example – there are only three types of measurement with which most people are normally concerned:

Length
Volume or capacity
Weight

Length

For short distances the *metre* will be used. A metre is a little longer than a yard (1 metre is actually 39.37 inches) but for many approximations it can be treated as roughly the same. 100 metres (or a *hecto*metre) is about 110 yards but if someone asks you how far it is to the Post Office it will make little difference if you say '200 yards' or '200 metres'; it is unlikely that you will have measured the distance exactly.

The metre is, for everyday purposes, divided into 100 *centi*metres. A glance at an ordinary ruler showing both inches and centimetres will show that about 30 centimetres (30.48 to be exact) is equal to 1 foot and that there are about 2½ centimetres in 1 inch. In other words, 5 centimetres is about equal to 2 inches – and 10 centimetres to 4.

For *approximations,* therefore, it will be sufficient to remember:

5 centimetres – 2 inches
10 centimetres – 4 inches

30 centimetres – 1 foot
9 metres – 10 yards
90 metres – 100 yards

As your own confidence in the metric system increases there is less need to convert and you will find yourself 'thinking metric'. Instead of thinking of your garage as being about 16 feet long, for example, you might perceive that it is about 5 metres. If you want to measure a window for curtains it will be easier to measure with a metric rule or tape than to measure in feet and inches and then convert the result and it will not be long before you think of the allowance for the hem as 15 centimetres rather than 6 inches.

What must be emphasised is that the introduction of metric measures need not (except, perhaps, as a matter of interest) lead to your making measurements which are not necessary now. If you do not know, and do not need to know, for example, the height of your dining table in inches you will not need to know it in centimetres.

If you buy a new bed you may find that the standard size is now a little larger than formerly. A single bed 200 centimetres long and 100 centimetres wide is about 3 inches longer and 3 inches wider than the older 6 ft 3 in × 3 ft bed. If you have very closely fitting sheets or the bed is built into a very tight recess you may notice the difference and, of course, twin beds will look odd unless they are both of one size. As the bedding industry comes into line with the furniture industry, the sizes of sheets, duvet covers and blankets are being adjusted to match the new bed sizes.

Longer distances

The measurement used in longer distances is the kilometre.* A kilometre is 1,000 metres and is equivalent to about 1,094 yards. 10 miles is about 16 kilometres, so the 30 mph speed limits will be converted to 50 km/h. Similarly, 100 kilometres per hour is about equal to 60 mph.

* Strictly speaking, the correct and logical pronunciation of this word lays stress on the first syllable – kílometre – but placing the stress on the second syllable – kilómetre – seems to be more common now. This pronunciation is almost invariably adopted by TV and radio commentators on athletics and motor racing.

A nautical mile (6,080 feet) is equal to 1.85 kilometres. A knot (1 nautical mile per hour) is therefore equal to 1.85 kilometres per hour.

Horse racing enthusiasts will be aware that 5 furlongs equal 1,100 yards. Since 1 yard, as already stated, is about $\frac{9}{10}$ metre, 5 furlongs can be regarded as the same as 1 kilometre for ordinary purposes. In fact, 1,100 yards is equal to 1,005 metres and, as mentioned above, 1,000 metres is equal to 1,094 yards.

Measures of length. Summary of approximate equivalents

1 centimetre	=	about $\frac{2}{5}$ inch
10 centimetres	=	about 4 inches
100 centimetres	=	1 metre = about 39 inches
1,000 metres	=	1 kilometre = about 5 furlongs

2 inches	=	about 5 centimetres
1 yard	=	about $\frac{9}{10}$ metre
10 miles	=	about 16 kilometres
30 miles	=	about 50 kilometres

There is a further sub-division of the metre which may be encountered – the millimetre (mm):

1,000 millimetres = 1 metre (10 mm = 1 cm)

It is equally correct to say that the length of this line is 2.5* ($2\frac{1}{2}$) centimetres or 25 millimetres.

0	1	2	centimetres
0	10	20	millimetres

It would be clumsy and incorrect to express it as 2 cm, 5 mm.

Volume or capacity

For small quantities, for which we use the terms fluid ounces, pints and gallons, the metric units are the millilitre and the

* If you are not quite sure how decimals work, turn to page 51 and read Chapter 12, *Decimals Are Not Difficult*, before you go any further.

litre. There are 1,000 millilitres in 1 litre (just as there are 1,000 millimetres in 1 metre).

The pharmaceutical industry was one of the first to change to metric measurement (no more drachms and minims for them) and most people are familiar with the 5 millilitre spoon (a little larger than a teaspoon) which is often used for medicinal doses. The millilitre is a very small quantity and will rarely be used by itself but bottles containing 100, 200, 250 or 300 millilitres will often be encountered.

It happens that 1 litre, which is equivalent to about 1¾ pints, is the space occupied by a cube with sides 10 cm long.

There are, therefore 10 × 10 × 10 cubic centimetres in 1 litre* (ie 1,000 cubic centimetres). It will be seen, therefore, that since there are 1,000 millilitres in a litre and 1,000 cubic centimetres in a litre, 1 millilitre is the same as 1 cubic centimetre and the contents of bottles, especially of cosmetics, will often be shown as, for example, 100 cc or 200 cc. This is exactly the same as 100 millilitres and 200 millilitres respectively.

Many people are already familiar with this without realising it. The internal size of the cylinders of cars is nowadays usually expressed in cubic centimetres (Fiesta 950, MG 1300, Rover 2000, for example) but sometimes in litres (Escort 1.3, Sierra 1.6, Montego 1.8 etc). 1300cc is the same as 1.3 litres.

Since there are 20 fluid ounces in a pint and this is equal to 568 millilitres, 1 fluid ounce is equal to 28.4 millilitres. For most purposes it will be convenient to regard 1 fluid ounce as equal to about 30 millilitres. For more detailed comparisons see pages 62 and 63.

Metrication in the bar

In bars, spirits are sold in quantities of ¼, ⅕ or ⅙ of a gill (a gill being 5 fluid ounces). One-sixth of a gill (known as 6 out) is the most common measure. It may be useful to think of 25 millilitres as about the quantity of a nip of gin, whisky or rum, though to be exact ⅙ gill = 23.7 ml.

Nearly everyone, having seen litre bottles of wines and spirits, has a very good idea of what a litre looks like. In

* If this is not clear, turn to the diagram on page 23.

supermarkets you may sometimes see milk sold in pints, half pints and litres on the same shelf. If the price of a pint is 28p the price of a litre should be about 50p.

For larger quantities, where gallons are now used, the litre will be the standard measure and it is useful to remember that 1 gallon is equal to about 4½ litres.

A kilderkin of beer contains 18 gallons. This is equal to about 81 litres. A 45 gallon barrel (of oil, for example) is about equal to 205 litres.

It must be remembered that, as in the case of length and distance, many of our daily activities are conducted in approximations. If we normally use one squirt of washing-up liquid to a bowlful of water or one bottle-cap of disinfectant to a bucket, it is not necessary to know the exact quantities involved either in fluid ounces or in millilitres. For many people, including some cooks, a knowledge of measures of capacity will be of no more than academic interest.

Centilitres and decilitres

There are two measures of capacity which you may sometimes come across, especially if you buy measuring jugs made on the Continent. Sometimes recipes will use these measures: the decilitre and the centilitre.

The contents of bottles of wines and spirits, apart from 1 litre and 2 litre bottles, are usually shown in centilitres (cl); 1 cl = 10 ml. The standard traditional bottles were ⅙ gallon (26⅔ fluid ounces), which is equivalent to 75.74 centilitres. It will be seen that twelve 26⅔ oz bottles (1 case) made 2 gallons (320 oz), but twelve 75 cl bottles make only 9 litres, equivalent to 317 ounces, about 1 per cent less. Brandy, liqueurs and some wines are often sold in 70 cl bottles; this is equal to 25 fluid ounces.

There are 100 centilitres in a litre, just as there are 100 centimetres in a metre. *Deci* means one-tenth (it is the same word-root as *decimal*) and a decilitre is one tenth of a litre. So we can see:

10 decilitres	=	1 litre
100 centilitres	=	1 litre
1000 millilitres	=	1 litre

Therefore:

10 centilitres	=	1 decilitre
100 millilitres	=	1 decilitre
and 10 millilitres	=	1 centilitre

Some wine goblets hold 6⅔ oz. This is a little less than 2 decilitres. A 7 oz tumbler would be almost exactly 2 dl.

Measures of weight

Of all the changes, weights are likely to cause most confusion, if only because not all unpackaged commodities will be sold in imperial quantities on one day and in metric quantities the next.

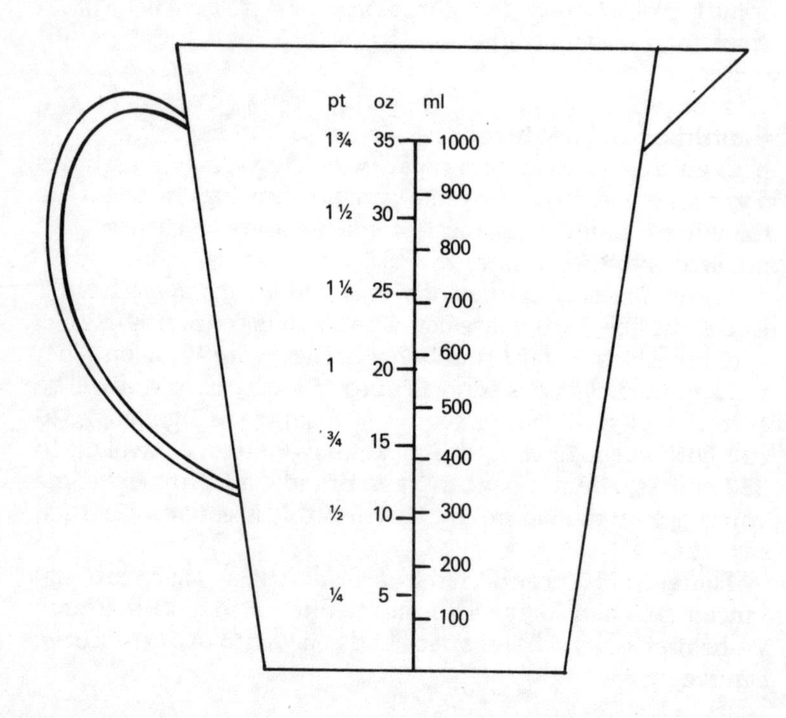

Measuring jug marked in ounces and millilitres

There will be no Metrication Day to correspond with the now almost forgotten D day when the price of nearly all goods changed from £sd to decimal currency. The butcher may decide to sell his products in metric quantities before or after the greengrocer and not all the shops in the High Street will change at once. All this is bound to lead to confusion, misunderstanding and accusations of profiteering.

In the case of packaged goods there need be little confusion. Tea, margarine, butter and sugar are, of course, normally sold in packages by weight and most buyers are now accustomed to the metric sizes – and the corresponding prices. If the change had taken place in a period when prices remained stable for many years it might have been much more difficult but shoppers are now accustomed to constant change in prices (usually upwards) and have become adaptable, or resigned.

If the price of a half-pound pack was 40p, the 250 g pack should cost 10 per cent more, ie 44p. 66p for 500 g is equivalent to 60p per lb.

There are some items which must, by law, be sold in specified quantities or multiples thereof, but for most commodities the only requirement is that the net contents should be specified and dual specification has already been introduced in most cases. Apart from trading standards officers and the most cost-conscious housewives, however, it will rarely be of great importance whether the contents of the breakfast cereal packet are stated to be 10 ounces or 285 grammes; the packet is recognised by its appearance and price.

One of the rare exceptions to the change from imperial weights and measures to metric is the proposed retention of troy weight, which is used for precious metals. 1 ounce troy (480 grains) is equal to 31.1035 grammes. There are 12 ounces troy in a pound but a pound troy is not the same as a pound avoirdupois.

Diamonds and pearls are measured in carats. 150 carats = 1 ounce troy. 1 diamond carat is about equal to 0.20 (1/5) gramme.

The base unit of weight in the SI* range is the kilogramme, comprising 1,000 grammes.

The kilogramme weighs about 2lb 3 oz. Many people travelling by air are familiar with the 20 kg baggage allowance,

* See page 8, Chapter 1, Metric and SI.

equivalent to 44 lb. For rough calculations it will be sufficient to remember that the half kilo (500 grammes) corresponds to the pound, though it is, in fact, about 10 per cent more.

Most people express their own weight in stones (14 lb units) and pounds (lb) though in the USA body weight is usually expressed in lb only. A 10 stone man or woman would be described as weighing 140 lb. This is equivalent to about 63 kg. A 16 stone heavyweight (224 lb) would scale just over 100 kg. It is useful to remember that a 50 kg sack is almost equal to 112 lb (1 cwt).

The tonne, or metric ton, contains 1,000 kg. This is equivalent to 2,205 lb. The imperial ton contains 2,240 lb, about 1½ per cent more. For many purposes it will not cause much inconvenience if they are treated as equal.

Weight. Summary of approximate equivalents

1 ounce	-	30 grammes
8 ounces	-	230 grammes
1 lb	-	450 grammes
1 stone	-	6.3 kilogrammes
1 cwt	-	50 kilogrammes
1 ton	-	1 tonne

In imperial measurements, confusion sometimes arises between fluid ounces and ounces avoirdupois. Since 1 gallon of water contains 160 fluid ounces and weighs 10 lb (160 oz) it follows that 1 fluid ounce of water weighs 1 ounce. One pint of water (⅛ of a gallon) weighs ⅛ of 160, ie 20 oz. In metric measurement the relationship is much simpler. One gramme is the weight of 1 cubic centimetre (1 millilitre) of water and 1 litre (1,000 cc) therefore weighs 1 kilogramme (1,000 grammes).

3. Writing It Down

It may seem inappropriate, in a book intended for the general reader, to insist on the importance of correct notation, but the International System (SI) does contain certain simple rules designed to avoid ambiguity and, for those unfamiliar with the symbols and encountering them for the first time, it will be no more difficult, and may prove to be easier, to adopt the correct style rather than a style of one's own.

The symbols (they cannot properly be called abbreviations) are as follows:

Linear measure

millimetre	–	mm
centimetre	–	cm
metre	–	m
kilometre	–	km

Capacity measure

litre	–	l*
decilitre	–	dl
centilitre	–	cl
millilitre	–	ml

Weight

gramme	–	g
kilogramme	–	kg
tonne	–	t

* Owing to the possibility of confusion with the figure 1 the use of the symbol '1' by itself will be discouraged and 'litre' will normally be written in full.

It is important to remember that in all cases:

1. Where small letters are used they should always be used: kg is correct; KG or Kg is not. The use of capital letters may lead to ambiguity since they may mean something quite different.
2. The plural is the same as the singular: 100 km is correct; 100 kms is not.
3. The symbols should not be followed by a full stop except at the end of a sentence.

The prefixes

It has been made clear that this book is concerned only with the measures which will be commonly encountered. It is worth remembering that the prefixes (eg kilo-, milli-) can be applied to *any* unit. Just as a millilitre (ml), is one-thousandth part of a litre and a millimetre (mm) one-thousandth part of a metre, so a milligramme (mg) is one-thousandth part of a gramme, though it is rarely used, apart from scientific purposes. There are other prefixes denoting each multiple of ten from one billion to one billionth part but apart from those already referred to none are likely to be encountered very often in everyday life.

Once more it should be emphasised that only one prefix should be used at once.

If we add 800 g and 700 g the sum can be shown as 1,500 g, or 1.5 kg, or even, in speech, as '1½ kilos' but never as 1 kg, 500 g.

Some other units

There are a number of other units, many of which are already in use in the United Kingdom, but are likely to present little difficulty. The watt (W)* the volt (V)* and the ampère (A)* are already in common use, though it is worth noting that the

* As mentioned above, most of the symbols will be spelt with lower-case (small) letters. The exception is that units derived from the name of a person (eg watt, volt, ampère, joule) are represented by symbols in upper case (capitals) – W, V, A, J, etc. The prefix (k, etc) remains in lower case. kW = kilowatt.

British Thermal Unit (BTu) from which the therm (used for measuring gas) is derived will eventually disappear. Since very few people have any clear idea of what these units actually represent – apart from calculations on the quarterly account – their disappearance will be almost unnoticed.

Time is unchanged in the metric system. The day, hour, minute and second remain the same the whole world over. For the scientist the system of dividing hours and minutes into sixtieths, dating back to the early Babylonians, may at times be inconvenient but the change to a decimal system of time would be much too great an upheaval of everyone's life for it to be even worthy of consideration.

Square measure

Most people are already familiar with the idea of square feet and square yards. If you have a carpet 15 feet × 9 feet you can say that it contains 135 square feet, or you could say that because it measures 5 yards × 3 yards it contains 15 square yards.

This is easy to see if we draw a diagram.

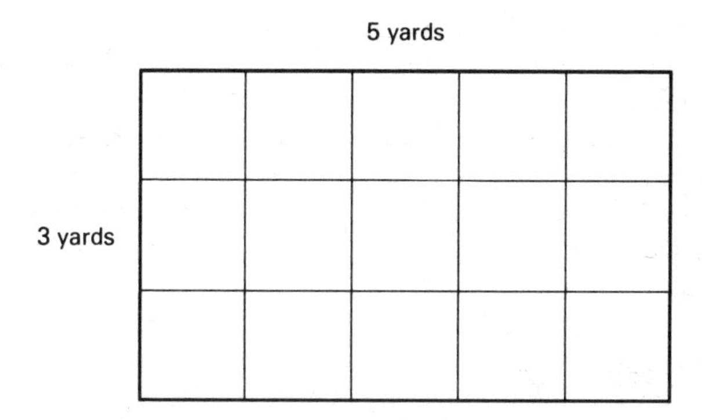

Exactly the same happens with metric amounts. 5 metres × 3 metres = 15 square metres. This is about 18 square yards.

There is one small difference in the way square measurements are written. Instead of 'square metre' you may sometimes come across the term m^2. This means exactly the same thing. Similarly cm^2 is a way of writing 'square centimetre'.

When we come to larger measures of area we shall probably find the *hectare* used. A hectare is 100 ares (pronounced to rhyme with 'fairs'). One are is 100 square metres but you will not often come across this measure. Since a hectare is 100 ares it contains 10,000 square metres, which is about $2\frac{1}{2}$ acres. If you picture a football pitch 110 yards long and 55 yards wide this has an area of 6,050 square yards (about $1\frac{1}{4}$ acres). In metric measurement this would be about 100 m × 50 m, which is equal to 5,000 m^2, which is, of course, half a hectare.

The goal area of a football pitch measures 6 yards × 20 yards. This is about 1.1 ares. A badminton court (20 feet × 44 feet) is about 0.9 ares. A tennis court (78 feet × 36 feet) is about 3 ares.

We do not often have to think about areas like this in everyday life but if you do ever come across reference to an are you can think of the size of a badminton court and if you want to imagine what a hectare looks like it may be useful to think of two football pitches put together.

Cubic measure

We have seen that m^2 means exactly the same as 'square metre'; cm^2 means 'square centimetre'. Cubic measures will often be written in a similar manner, ie as m^3 or cm^3, meaning cubic metres or cubic centimetres.

If you were a gardener and you wanted to spread a covering of soil 2 inches thick over an area 60 ft × 30 ft you could calculate that this would need 300 cubic feet, or about 11 cubic yards. In the same way if you wanted to spread it 5 cm thick over an area 18 m × 9 m you could multiply 18 × 9 × 0.05 to arrive at a figure of 8.1 m^3.

On the other hand, if you never do calculations like this in feet and inches you will not have to do them in metres and centimetres. It just happens that the metric calculations are usually easier if you should have to do them.

A point worth noting is that whereas we have to use a formula to calculate the relationship between cubic yards and gallons or cubic feet and pints it is much easier in metric because we know that 1 litre is 1,000 cc so that 1,000 litres = 1 m^3. This is easier to understand if we think of a litre as a box with sides 10 cm × 10 cm × 10 cm. You would need 1,000 such boxes to fill a space 100 cm × 100 cm × 100 cm, which is of course 1 m × 1 m × 1 m, or 1 m^3.

There are 100 one-litre boxes on each layer and there are 10 layers. Therefore it is seen that 1000 litres = 1 m^3.

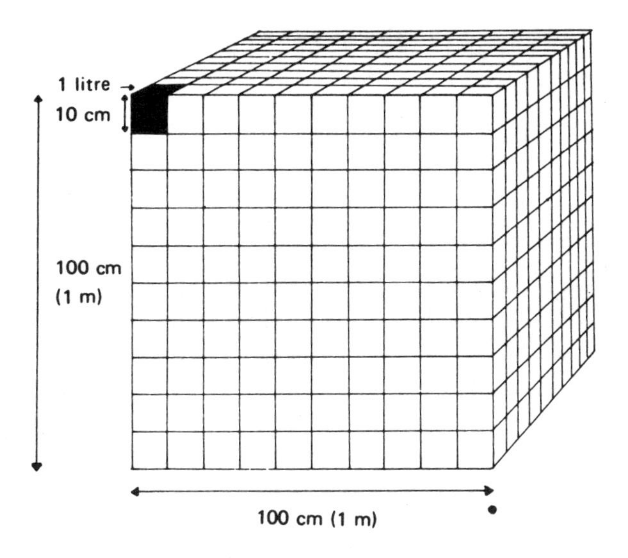

We have already seen that 1 cc of water weighs 1 gramme, so 1,000 cc (1 litre) weighs 1 kilogramme. It follows that 1,000 litres (1 m^3) weighs 1,000 kilogrammes (1 tonne). This only applies to water. Other substances have a different density and it might require more or less in cubic measurement to make 1 tonne in weight.

prefix	**symbol**	**factor**	**example**	
mega-	M	million	MW	megawatt*
kilo-	k	thousand	kg	kilogramme
hecto-	h	hundred	ha	hectare
deci-	d	tenth	dl	decilitre
centi-	c	hundredth	cm	centimetre
milli-	m	thousandth	mm	millimetre

There are also prefixes for larger (up to million million million) and smaller (down to million million millionth) factors but these will rarely be encountered except in scientific or technical use.

* One megawatt (1 million watts) is equal to 1,000 kilowatts.

4. Temperature

Temperature should now be measured in the Celsius scale, which many people know, and will continue to refer to, as centigrade. Unless you happen to be a heating engineer or a refrigerator manufacturer or your work involves calculations of precise temperature variations, it is unlikely that the change from Fahrenheit to Celsius will have caused you any problems. Those watching weather forecasts on TV or hearing them on the radio will now be accustomed to the use of Celsius, though the Fahrenheit equivalent is usually given also. Hospital nurses now measure your body temperature with a metric thermometer and if you are not ill it will probably show about 37 degrees, which is normal body temperature. The main thing to keep in mind is that nothing other than the scale for measuring temperature has changed; you will need to wrap up on a cold day whatever kind of thermometer you use.

One of the mental blocks which a lot of people suffer from when the question of changing from Fahrenheit to Celsius (or centigrade) comes up is concerned with conversion. Most people recollect having been told at some time in their schooldays what the formula was but have now forgotten it, or remember it imperfectly. The trouble with most of these easy methods of conversion is that if you do not get the formula exactly right they are not of much use.*

It is often just as easy, if you have to convert, to refer to a conversion table. However, by far the best way to overcome the problem is to train yourself to think in the new system. An easy way of doing this is to associate a few temperatures in degrees Celsius (centigrade) with things that you are familiar with. For example, you could use four Celsius temperatures as your guide to all others. These could be:

1. Freezing point of water (32 °F – 0 °C)

* The correct formula appears on page 26.

2. Normal room temperature (about 68 °F – 20 °C)
3. Body temperature (98.4 °F – 36.9 °C)
4. Boiling point of water (212 °F – 100 °C)

Having decided on a few anchor points like these, you should then force yourself to think about the temperature in degrees Celsius whenever you say, or someone else says, 'It's *cold*' or 'It's *hot*' or 'It's *freezing*', or even 'Isn't it *warm* for this time of the year.' This is the only way you will learn to think in metric temperatures. To assist you in the selection of your anchor

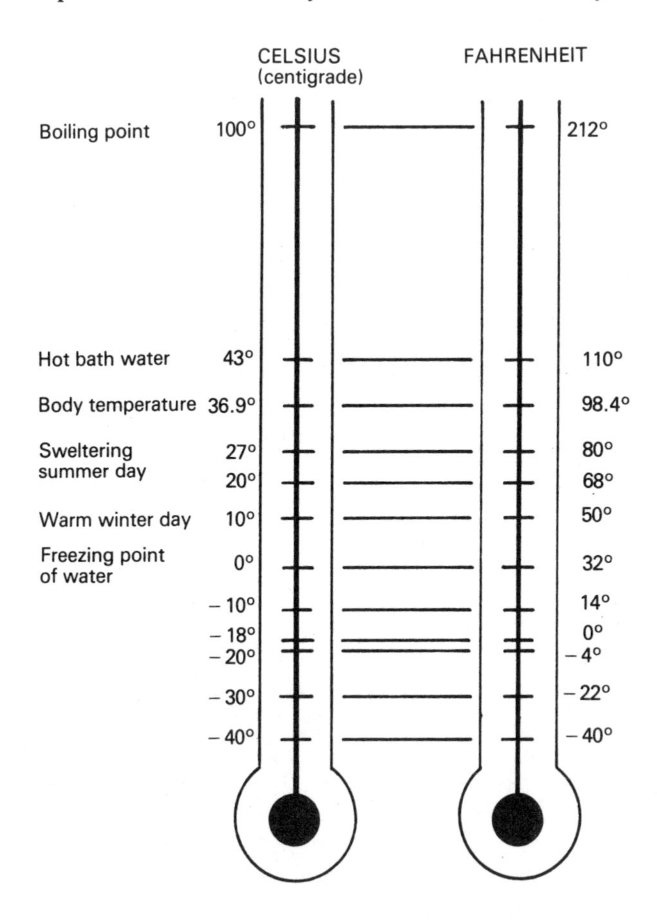

Approximate comparisons between the Celsius and Fahrenheit thermometer

points the chart on page 25 gives a few examples to choose from.

For those who cannot be bothered to adopt new ways of thinking centigrade, a conversion table will be found on page 28. The graph on page 27 allows you to see at a glance the approximate equivalent of any temperature whether in centigrade or Fahrenheit.

Just in case you did want to use the formula for conversion it is:

C to F: Multiply by $^9/_5$ and add 32.
F to C: Subtract 32. Then multiply by $^5/_9$

eg 25 °C = $^9/_5 \times 25 + 32 = 45 + 32 = 77$ °F
68 °F = $(68 - 32) \times {}^5/_9 = 36 \times {}^5/_9 = 20$ °C.

The fraction $^5/_9$ is derived from the fact that the differences between the freezing and boiling points of water are 100 °C and 180 °F (212 – 32). $^{100}/_{180} = {}^5/_9$

There is another formula which you might find easier. Because minus 40 °C = minus 40 °F you can:

ADD 40 to the temperature (either C or F)
MULTIPLY by $^5/_9$ (F to C) or $^9/_5$ (C to F)
SUBTRACT 40 from the answer.

eg 25 °C + 40 = 65
$65 \times {}^9/_5 = 117$
117 – 40 = 77 °F.

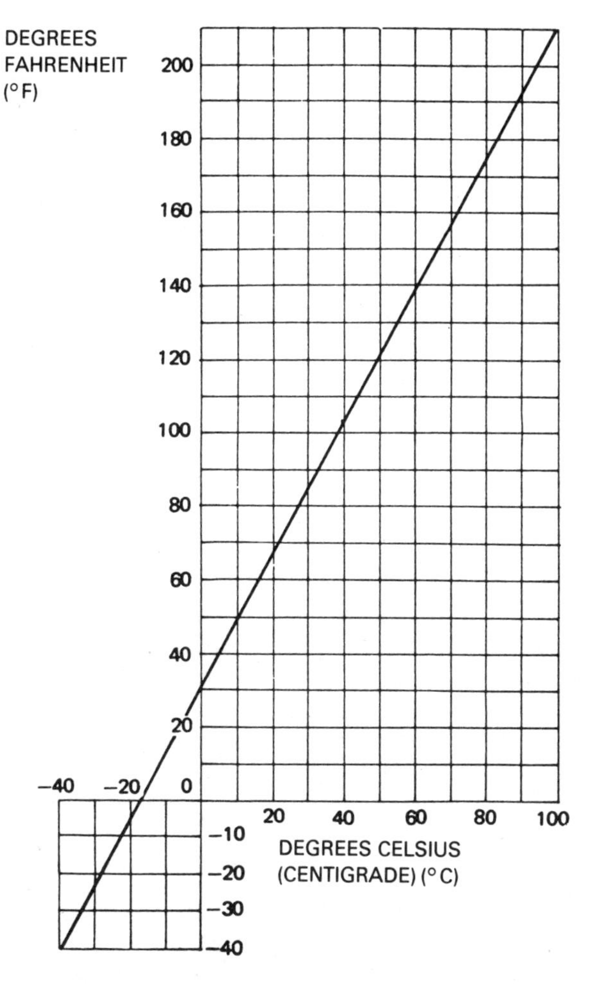

To use this chart look at the temperature on the scale marked F or C as the case may be. Draw a line at right angles. Where this line cuts the diagonal will be exactly opposite to the temperature conversion.

Conversion Table

Celsius °C	*Fahrenheit* °F	*Fahrenheit* °F	*Celsius* °C
–40	–40	–40	–40
–35	–31	–35	–37.2
–30	–22	–30	–34.4
–25	–13	–25	–31.7
–20	–4	–20	–29.0
–15	5	–15	–26.1
–10	14	–10	–23.3
–5	23	–5	–20.6
0	32	0	–17.8
5	41	5	–15.0
10	50	10	–12.2
15	59	15	–9.4
20	68	20	–6.7
25	77	25	–3.9
30	86	30	–1.1
35	95	35	1.7
40	104	40	4.4
45	113	45	7.2
50	122	50	10.0
55	131	55	12.8
60	140	60	15.6
65	149	65	18.3
70	158	70	21.1
75	167	75	23.9
80	176	80	26.7
85	185	85	29.4
90	194	90	32.2
95	203	95	35.0
100	212	100	37.8
		110	43.3
		120	48.9
		130	54.4
		140	60.0
		150	65.6
		160	71.1
		170	77.7
		180	83.3
		190	88.8
		200	93.3
		210	98.9
		212	100

5. Metrication in Sport

Sports fans need not worry that their favourite team might be placed at a disadvantage when the size of the playing area is measured in metres instead of yards. Football and cricket pitches will stay the same size whatever measurement is used. Even in horse racing the changes, if any, will be quite small.

Horse racing

At present the distances of races are not always exact multiples of a furlong. The Vaux Gold Tankard at Redcar is run over 1 mile, 6 furlongs and 132 yards, the same distance as the Doncaster St Leger. The Chester Cup distance is 2 miles, 517 yards. The Royal Hunt Cup at Ascot was run over 7 f 166 yards before 1930, 7 f 155 yards from 1930 to 1954 and has been exactly a mile only since 1955. On the Continent 1,400 metre races are treated for comparison purposes as if they were 7 furlongs; 1,600 m is equivalent to 1 mile.

The difference is only a few yards and in any case there can be no exact comparison of times between races run at different places on different days owing to variations in the conformity of the courses and the state of the going.

Jockeys' weights are expressed in stones and lb. When these are changed to kilogrammes (which will probably not be for some time yet) it will be much easier to see the difference between, say, 49 kg and 53 kg (ie 4 kg) than that between 7 st 10 lb and 8 st 5lb (9 lb).

Athletics

Those who follow athletics will already be familiar with distances measured in metres. In many cases the differences are insignificant. Two hundred and twenty yards is about 4 feet more than 200 metres. New tracks are being built to international specification and all tracks in Britain will eventually be converted, but meanwhile events measured in yards are still occasionally being run.

Television viewers may have noticed that in field events commentators usually use feet and inches (in high jump and long jump, for example) even though the measurements are made in centimetres. To the uninitiated a high jump of 226 centimetres is not much less meaningful than 7 ft 5 in. To those who would find it difficult to scale 4 feet it just seems enormously high. What is important is that it is approaching the world record, whatever method of measurement is used.

As far as track events are concerned there can be no exact comparison between even slightly different distances owing to the effect of distance on stamina. 9.1 seconds for 100 yards is about equivalent to 10.0 seconds for 100 metres but it is not the same thing. Times for 880 yards and 800 metres will be a fraction of a second different. Because exact comparison of races run over different distances is not possible, records set up over distances measured in yards will stand for all time now that these distances are no longer run in international events.

Cricket

The length of a cricket pitch is 22 yards (one chain). Although it might be more convenient to measure in metric it is unlikely that traditionalists will ever allow this to be changed to 20 metres, which is 11.5 cm (about 4½ inches) less.

6. Metrication in DIY

Do-it-yourself enthusiasts are already meeting some of the difficulties in the transitional period which the change to metric measurements will bring for everyone.

Some things, such as netting staples or upholstery pins, are usually found pre-packed and will therefore present no problem. In DIY stores nails are also pre-packed but the more old-fashioned hardware shops continue to sell nails, especially the larger sizes (eg 100 mm and 150 mm) by weight and still measure them by the pound. Packets of nails may be in round amounts (eg 40 g) but sometimes in what might seem to be odd weights such as 198 g. This is equal to 7 oz. Screws are usually sold by number, in multiples of 10, except when they are pre-packed.

Not all manufacturers, wholesalers and retailers are converting at the same time; old stocks often have to be cleared; the retailer may wish to pretend that there has been no change and refer to the new sizes as if they were the old ones.

If you use particular items regularly it might be as well to ask your supplier what his programme is for the change to metric. He will have to change sooner or later.

In most cases the metric sizes will be so close to the old sizes that it will make little difference unless you work to very precise limits. Sometimes the change to metric will be used as an opportunity for the manufacturer to rationalise sizes so that some sizes may be discontinued altogether.

If you are buying timber you may find that your supplier has the metric sizes but still sells by the foot. This is only an attempt to ignore or delay the change.

Standard wood sizes in metric are very close to the sizes we have been used to. 4″ × 2″ is now 100 mm × 50 mm. 2″ × 2″ is 50 mm × 50 mm. 2″ × 1″ is 50 mm × 25 mm. If you mark out these sizes on a piece of paper you will see that there is very little difference (remember that 50 mm = 5 cm).

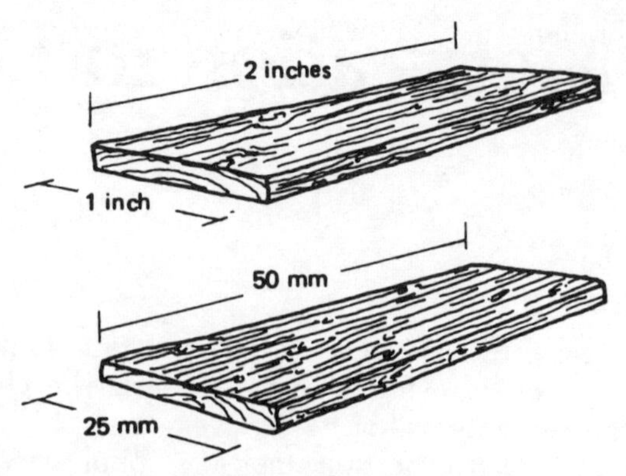

Standard metric widths and thicknesses of timber, with approximate equivalents:

mm	*in*
12	½
16	⅝
19	¾
22	⅞
25	1
32	1¼
38	1½
50	2
63	2½
75	3
100	4
125	5

Then in 25 mm stages to 300 mm (12 inches)

Boards are still made in the imperial sizes formerly used but may be described in millimetres (eg 2,440 × 1,220, equivalent to 8 ft × 4 ft).

Glass is graded by thickness, from 2 mm (suitable for picture frames) to 12 mm (plate glass).

Paint is sold in tins of 250 ml and 500 ml (roughly equivalent to ½ pint and 1 pint), 1, 2.5 and 5 litres.

If you wanted to calculate the amount of paint needed for a job you could measure it up, using a metric rule or tape, to get the area in square metres. This is done by multiplying the height and length of each wall and adding the answers together. For example:

240 cm × 300 cm	=	72,000 cm^2
240 cm × 300 cm	=	72,000 cm^2
240 cm × 210 cm	=	50,400 cm^2
240 cm × 210 cm	=	50,400 cm^2
Total		244,800 cm^2

To convert this to m^2 you divide by 10,000 (move the decimal point 4 places to the left) and you have 24.48 m^2 (24.5 m^2 to one decimal place). Some paints show on the can the area that can be covered by a litre. If not you can ask the retailer how much paint you need for about 25 square metres.

The same kind of calculation would be used for measuring a room for a carpet. You can either measure in metres and decimals of a metre or you can measure in centimetres as indicated above and divide by 10,000 (100 × 100). For example:

5.90 m × 4.75 m	=	28.025 m^2
or 590 cm × 475 cm	=	280,250 cm^2

The main danger in doing calculations like this is in misplacing the decimal point, but you can always make a rough check. If you came up with a figure of 2.44 m^2 or 244 m^2 for the walls a glance would show you that there must be a mistake.

Once you have got used to it you will find that measuring in metric is easier in many ways than measuring in feet and inches. The methods you use are unchanged. It is only the figures that are different.

Plumbing pipes and fittings are now made in metric sizes. The older sizes may still be obtainable but they are becoming increasingly rare. Although the metric sizes are sometimes very close to the imperial they are not entirely compatible. If it is necessary to fit a metric component in an older system it is possible to obtain adaptors to connect ½ inch to 15 mm, ¾ inch to 22 mm and so on.

7. Metrication and the Motor Car

The most striking change affecting motorists in recent years is the introduction of metric pricing for petrol. A few garages still sell petrol in gallons but service stations are no longer required by law to display the gallon price on the forecourt.

For practical purposes 4½ litres can be treated as equal to one gallon. The true figure is 4.546, so that:

37p per litre = 168.2p per gallon
38p per litre = 172.7p per gallon
39p per litre = 188.3p per gallon
40p per litre = 181.8p per gallon

To do the calculation in reverse, multiply the gallon price by 0.22:

£1.78 per gallon = 39.18p per litre
£1.79 per gallon = 39.38p per litre
£1.80 per gallon = 39.60p per litre

For those who habitually buy £5 or £10 worth of petrol these figures may be of no more than academic interest.

The same conversion factors can be used to calculate petrol consumption:

30 miles per gallon = 6.6 miles (10.56 km) per litre
40 miles per gallon = 8.8 miles (14.08 km) per litre

The chart on the opposite page makes it easy to make detailed comparisons.

On the Continent it is more usual to measure petrol consumption in litres per 100 km. Thus:

10.56 km per litre = 9.5 litres per 100 km
14.08 km per litre = 7.1 litres per 100 km

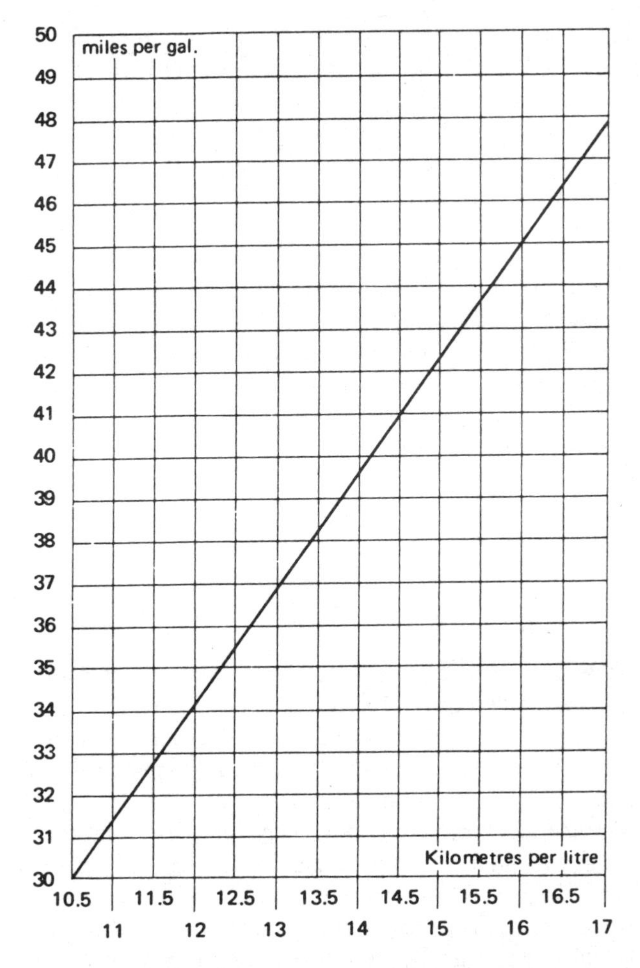

To read this chart, take a figure of miles per gallon on the vertical axis, draw a line across the graph and read from the horizontal axis the equivalent figure in km per litre where this line cuts the diagonal line.

Example: 41 mpg = 14.5 km/litre

Although British governments, especially the present one, have always been reluctant to face the need to change our speed limits and signposting from miles to kilometres it now seems that the change cannot be resisted much longer, despite arguments against it on both sentimental and practical grounds. It is likely that, in the next few years, much more dual marking of signs will be introduced but it will be a long time before every map and guide has been revised and reprinted.

An EC requirement to harmonise regulations limiting the speeds of heavy-goods and passenger service vehicles will require new limits of 50 km/h and 100 km/h in various circumstances. This is likely to necessitate the erection of traffic signs showing these speeds or their equivalents in miles per hour. There would inevitably be some confusion with existing signs showing general limits but this could be resolved either by converting our present speed limits to the nearest round metric equivalent or by amending these limits to 31 mph instead of 30 mph and 62 mph instead of 60.

Whichever expedient is adopted, it is beginning to look as if the mile is doomed as well as most other imperial measures but beer drinkers are likely to continue to insist on retaining measures of 568 ml and to refer to them as pints.

Tyres

Tyre sizes are usually expressed in a mixture of metric and imperial measurements. 145SR12 means a section width of 145mm and a measurement of 12 inches diameter from rim to rim. Entirely in metric this will eventually become 145/305; similarly, 155SR13 will become 155/330.

In service station forecourts tyre pressure gauges often show pressure in *bars*. The bar is a metric unit of atmospheric pressure. We are all familiar with weather forecasters' announcements that today's atmospheric pressure is (say) 1,000 millibars (1 bar). For most cars the normal tyre pressures should be approximately 2 bars, which is equivalent to 28½ lb per square inch but the operating manual should be consulted to give the exact figure, which will vary according to the load carried and the driving conditions.

Feeler gauges

Feeler gauges may be measured in thousandths of an inch (thou) or in millimetres (mm). Since 1 millimetre is 0.03937 inch it is safe to treat 40 thou as equal to 1 millimetre. This would be the average spark-plug gap for lean-burn engines.

Conventional engines burning leaded fuel require a slightly smaller gap (perhaps 30 thou = 0.75 mm).

Nuts and bolts

Most foreign cars have nuts and bolts in metric sizes but on some cars made in Britain imperial sizes may be found. $\frac{3}{4}$ inch is almost exactly equal to 19 mm (there are 25.4 mm in 1 inch) but other sizes are not compatible.

The table below gives approximate equivalents which could be used if the correct socket wrench were not available but this should not be resorted to except in emergency since damage might be done to the nut if a wrench of incorrect size is used. For some reason known only to the manufacturers, most oil filters have non-standard threads. Unless the part is supplied by the manufacturer of the car for which it is required it may be found that the thread does not fit, even though the filter is similar in size and design.

Approximate equivalents of thread sizes

inches	*millimetres*
$\frac{5}{16}$	8
$\frac{3}{8}$	10
$\frac{7}{16}$	11
$\frac{1}{2}$	13
$\frac{9}{16}$	14
$\frac{5}{8}$	16
$\frac{11}{16}$	17
$\frac{3}{4}$	19

8. Metrication in the Kitchen

Most recipe books published in recent years show quantities of ingredients in both imperial and metric. Provided you have suitable measuring spoons and jugs, this should present no difficulty.

Some people rely on approximation, especially for well-tried recipes. For these people, since there is no exact measurement in ounces or fluid ounces, there will be no need to measure in grammes or millilitres.

When it is necessary to convert a recipe from imperial to metric it is important to remember that recipes are not like pharmaceutical prescriptions. The quantities specified are not minutely exact and sometimes vary from one cookbook to another. In any case, not many kitchens are equipped to measure very exactly, either in ounces or in grammes. In commercial catering, where the recipe has been carefully costed and the portion size is standardised, there may be little scope for the exercise of discretion but in the home the exact quantities of ingredients used are not critical provided the relative proportions of the various ingredients are correct.

One ounce in weight is 28.4 grammes (28.352 to be precise) and 1 fluid ounce is 28.4 millilitres. One cubic centimetre of water weighs 1 gramme and 1 litre weighs 1 kilogramme. In converting recipes it will usually be found convenient to use approximate equivalents, either 1 oz = 25 g (1 pint = 500 ml) or, preferably, 1 oz = 30 g (1 pint = 600 ml). What is essential is to maintain the *balance* of the ingredients. To take a simple example:

Pancake batter

½ pt (10 fl oz) milk
4 oz flour
1 large or 2 small eggs; pinch of salt.

Here the quantity of milk is 2½ times that of flour.* Apart from the difference in the total quantity – slightly less or slightly more than the original recipe – it does not matter whether you use 250 ml milk and 100 g flour (with a small egg) or 300 ml and 120 g (with a large egg). It would, of course, be quite wrong to mix the two approximate equivalents using, say, 250 ml milk and 120 g flour or to mix imperial and metric measures, eg ½ pint milk and 100 g flour.

Here is another example of the two methods of conversion. Which you use is a matter of choice but 1 oz = 30 g is nearer to the original:

Bechamel sauce

Milk	½ pint	250 ml	or	300 ml
Cream	2 oz	50 ml	or	60 ml
Butter	1 oz	25 g	or	30 g
Flour	1 oz	25 g	or	30 g

Plus shallot, bayleaf, peppercorns, etc.

In each case the proportions are: milk 10: cream 2: butter 1: flour 1.

For those with no metric measuring equipment a medicine spoon (5 ml), as supplied free with some prescriptions, may be found useful for small quantities. For larger quantities the following may be used as a rough guide:

1 level tablespoon (20 ml)	=	10 g flour	=	15 g sugar
1 teacupful (200 ml)	=	100 g flour	=	150 g sugar

If you are accustomed to using recipes which specify only cupfuls or spoonfuls you need not worry about conversion to metric amounts. You will just go on using the same cups and spoons as you did before. If you come across decilitres or centilitres in recipes the comparison chart below may be found useful. Remember, though, that the conversions shown are not exact. If you do require exact conversion, for this or any other purpose, refer to the conversion table on page 62.

* A scientist may tell you that this is not strictly correct since the specific gravity of milk is not the same as that of water but the difference is so small as to be immaterial for practical purposes.

Comparison chart

millilitres	=	*centilitres*	=	*decilitres*	=	*approximately*
4 ml	=	0.4 cl	=	0.04 dl	=	one average teaspoonful
5 ml	=	0.5 cl	=	0.05 dl	=	one medicinal spoonful
10 ml	=	1 cl	=	0.1 dl	=	one dessertspoonful
25 ml	=	2.5 cl	=	0.25 dl	=	one tot of gin, whisky, etc
30 ml	=	3 cl	=	0.3 dl	=	one fluid ounce (1 fl oz)
60 ml	=	6 cl	=	0.6 dl	=	one small glass of sherry
140 ml	=	14 cl	=	1.4 dl	=	quarter pint (¼ pt)
290 ml	=	29 cl	=	2.9 dl	=	half pint (½ pt)
580 ml	=	58 cl	=	5.8 dl	=	one pint (1 pt)
1000 ml	=	100 cl	=	10 dl	=	one litre

One litre = approximately 1¾ pints

When measuring butter, lard, etc from a 250 g packet it will be found most convenient to divide the packet into five and then, if necessary, divide one of these sections by two to give 25 g. If you find the 30 g equivalent of an ounce more convenient you will not go far wrong if you divide a 250 g packet by four, to give 62.5 g and then again to give a 31.25 g portion.

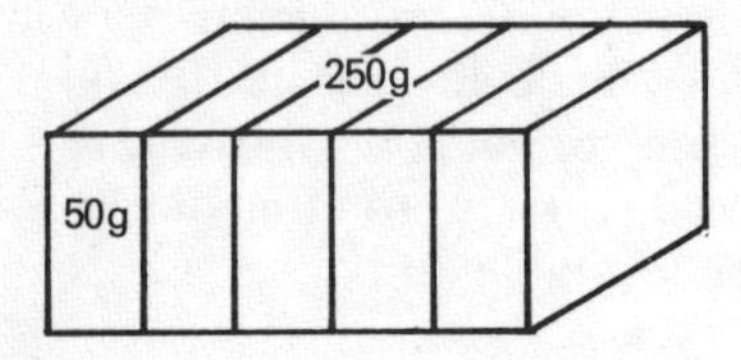

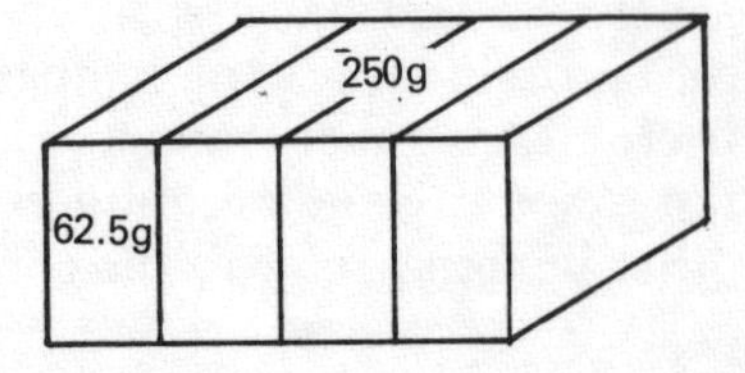

Where recipes give cooking temperatures the following table can be used:

	°C	°F	Gas mark
	140	285	1
Slow	150	300	2
	170	340	3
Moderate	180	355	4
	190	375	5
Mod hot	200	390	6
Very hot	220	430	7
	230	450	8

For more information about Celsius and Fahrenheit temperatures turn to page 24.

9. Clothing Sizes

In metric the basic unit of measurement for ready-made clothing is the centimetre. Fractions of a centimetre are not used and graduations of sizes are often in steps of 4 or 5 centimetres. Conversions are not, therefore, exact and it will be found most satisfactory to ascertain your metric sizes for various items and remember them.

Some clothing sizes vary between Britain and the United States but the metric sizes are universal.

Women's dresses and skirts

		Bust		Hips		Waist	
British	*US*	*cm*	*in*	*cm*	*in*	*cm*	*in*
8	8	76	30	81	32	58	23
10	8	81	32	86	34	58	23
12	10	86	34	91	36	61	24
14	12	91	36	97	38	66	26
16	14	97	38	102	40	71	28
18	16	102	40	107	42	76	30
20	18	107	42	112	44	81	32

Children's clothing

Children's clothes may be marked in centimetres *and* inches. Up to the age of 12 months the size is related mainly to the weight of the child. Thereafter, sizes relate to height; approximate ages are sometimes indicated also.

	Height		Weight		Chest	
Age	*in*	*cm*	*lb*	*kg*	*in*	*cm*
3 months	24½	62	12	5.5	17¾	45
6 months	27	68	18	8	18½	46
9 months	29	74	21	9.5	18⅞	48
12 months	31½	80	24	11	19⅝	50
18 months	34	86			20¼	52
2 years	36	92			20⅞	53
3 years	38	98			21½	55
4 years	40	104			22⅛	57

Men's clothing

Metric chest measurements for jackets, cardigans, pullovers, etc and waist/inside leg measurements are in centimetres. Shirt sizes show the circumference of the neckband in centimetres:

Inches	14½	15	15½	16	16½	17
Centimetres	37	38	39–40	40–41	42	43

Hats

The size of hats at present is the width (in inches) of the inside of the hat. In metric the size is the circumference (in centimetres) of the head. Since the inside of a hat is not circular there is no exact ratio of inch sizes to metric. For a time, dual marking, already common, will continue (eg 7⅜–60).

Inches	6½	6⅝	6¾	6⅞	7	7¼	7⅜
Centimetres	53	54	55	56	57	58	60

Footwear

Women's shoes

Metric shoe sizes are based on a combination of length and width:

British	4½	5	5½	6	6½	7
US	6	6½	7	7½	8	8½
Metric	38	38	39	39	40	41

Men's shoes

British	7	7½	8½	9½	10½	11
US	8	8½	9½	10½	11½	12
Metric	41	42	43	44	45	46

Women's hosiery

British	8	8½	9	9½	10	10½
Continental	0	1	2	3	4	5

Men's socks

Inches	9½	10	10½	11	11½
Centimetres	38–39	39–40	40–41	41–42	42–43

10. International Paper Sizes

The international paper sizes now commonly used in Britain are not new but there is still some confusion as to why they were introduced to replace the traditional sizes and what useful qualities the new system possesses.

In fact, British Standard 730 was published in 1937, but it was not until 1959 that the recommendations of the International Organisation for Standardisation were accepted in the United Kingdom.

The sizes formerly in use – and still used for special purposes – are based on subdivisions of large sheets of paper of various dimensions. *Folio* means folded in half, *quarto* (4to) folded into four, *octavo* (8vo) folded into eight, and so on. Crown paper is 20 × 15 inches; double Crown is 30 × 20 ; quad Crown is 40 × 30. A Crown octavo book would have pages one-eighth of Crown size, ie 7½ × 5 inches (7½ being half of 15 and 5 a quarter of 20).

Similarly, Royal is 25 × 20 inches and Royal 8vo is 10 × 6¼. There are some 20 basic sizes of paper used for printing and a different series of 25 sizes for drawing and writing paper ranging from Emperor (72 × 48 inches) to Post (15 × 12½) and each of these could be folded into halves, quarters, etc.

The International System is based on a different principle. The basis is a rectangle having an area of one square metre and sides in the proportions 1:√2.* This gives dimensions of 841 × 1,189 millimetres. If you multiply these two numbers you will see that the area is 999,949 mm^2, ie almost exactly 1,000,000 square millimetres, or one square metre. This size, equivalent to approximately 33 × 47 inches, is known as Ao. Sizes larger than Ao are preceded by a figure. Thus, 2A means twice the

* The square root (√) of a number is the number which, multiplied by itself, produces the original number. Thus, 2 is the square root of 4; 4 is the square root of 16; 3 is the square root of 9; 10 is the square root of 100.

size of Ao (but in the same proportions); 4A is four times the size, ie 1,682 mm × 2,378 mm.

Half Ao (ie 841 mm × 594 mm) is A1; half of that is A2, half of that is A3, and so on. These are trimmed sizes. The most common size in commercial use is A4 (210 mm × 297 mm – 8.27″ × 11.69″), which is slightly larger than the quarto (8″ × 10″) formerly used but smaller than foolscap folio (8½″ × 13½).

The great advantage of the A sizes is that, since each rectangle, regardless of its size, is in the same proportions, (approximately 14:10) photographic enlargement or reduction is greatly simplified. When enlargement takes place the shorter side is doubled and the longer side becomes the shorter. In reduction the longer side is halved; the shorter side is unchanged but becomes the longer. Thus, when A4 becomes A3, the 210 mm side becomes 420 mm but the 297 mm side is unchanged. When A4 becomes A5 the 210 mm is unchanged but the 297 mm is halved. This is clearly seen in the table below.

'A' sizes

Designation	mm
A0	841 × 1,189
A1	594 × 841
A2	420 × 594
A3	297 × 420
A4	210 × 297
A5	148 × 210
A6	105 × 148
A7	74 × 105

There are other subsidiary series. 'B' sizes have been devised for use in special cases where sizes between two 'A' sizes are required:

'B' sizes

Designation	mm	Designation	mm
Bo	1,000 × 1,414	B4	250 × 353
B1	707 × 1,000	B5	176 × 250
B2	500 × 707	B6	125 × 176
B3	353 × 500	B7	88 × 125

The 'C' series is used for envelopes. These are designed to contain 'A' size pieces of paper folded once or twice. Thus, for example, C6 (114 mm × 162 mm) accommodates A4 sheets folded twice:

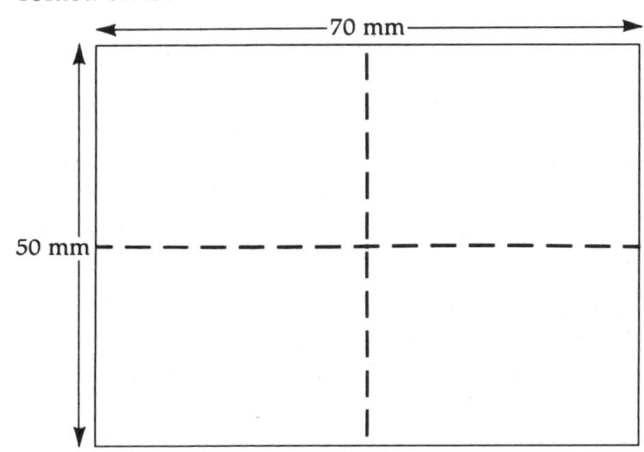

or A5 sheets folded once:

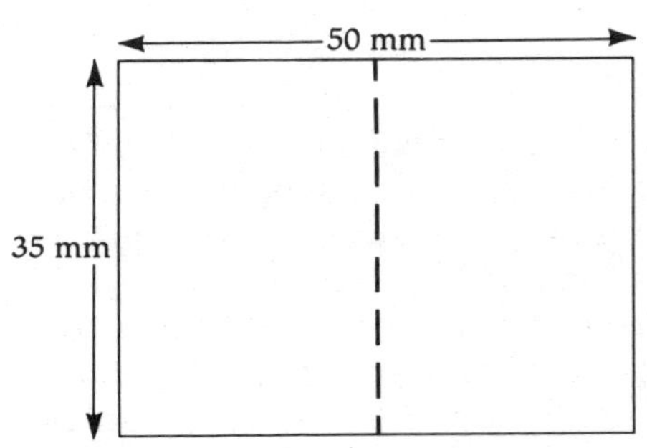

A C6 envelope would, of course accommodate A6 without folding.

11. Thinking Metric

Conversion tables have been included in this book for convenience but the use of such tables is not the best way of coping with the change. The easiest and least time-consuming way is to 'think metric'. This is easier to talk about than to do but it is worth while to try.

One exercise which has been found useful is known as 'metricating yourself'. All you have to do is to measure a number of features associated with yourself and then memorise them in metric quantities. Then when a situation arises where you need to estimate the size of an object, or a distance, you can ask yourself how it compares with a size or distance you already know.

If you are, say, 5 feet 7 inches tall you have a good idea of how high other people and things are by comparison. Exactly the same method will be used when you think of your height as 1.70 metres. If the span from your outstretched thumb to your little finger is 8 inches it will be just as easy to think of it as 20 cm.

If you persevere with this you will discover quite soon that you do not need to rely on the things you originally memorised. You will be adopting metric concepts in much the same way as you soon abandon the phrase book or dictionary if you live for a time in a foreign country. If you are in France when you hear a clock striking you may think at first 'One, two, three, four – that must be quatre heures' but before long you will find it coming into your mind as 'Un, deux, trois, quatre'.

If you want to try this method fill in the answers to the questions which follow. You need not try to do them in your head at this stage. Use a metric tape measure or weighing machine or, if there is not one handy, measure first in imperial measures and write down the metric conversions.

Metricating yourself

1. What is your own weight? .... kg
2. What is your own height? .... m
3. With your hand outstretched what is the distance from your thumb to your little finger? .... cm
4. What is the width of your thumb? .... mm
5. What size is the smallest square in which you can stand? .... cm^2

If you own a car:

6. What is the cubic capacity of the engine? .... cc (cm^3)
7. What are the car's overall measurements?
   length .... m
   width .... cm
   height .... cm
8. What are the correct tyre pressures? .... bars

Your home:

9. What are the dimensions of your front door?
   height .... m
   width .... cm
10. What is the length of your back garden? .... m
11. What is its area? .... ares
12. What is the distance from your home to the station/shops? .... km
13. What is your average speed when walking from your home to the station/shops?

$$\frac{\text{Distance covered} \ldots\ldots\ldots\ldots \text{m}}{\text{Time taken} \quad \ldots\ldots\text{minutes}} \times \frac{60}{1000} = \ldots\ldots\text{km/hour}$$

14. What is the cubic capacity of your bedroom?
Height .... m × width .... m
× Length .... m = ... m^3

15. At what temperature do you feel comfortable
(a) Working? .... °C
(b) Relaxing? .... °C

16. At what temperature do you feel uncomfortably hot?
.... °C

If you can memorise these few items and relate them to anything else you meet you will be well on the way to metricating yourself. It is worth trying since it will save you a lot of trouble in the long run.

12. Decimals Are Not Difficult

There is really no difficulty about understanding decimals once the basic idea has been grasped. It is because many people imagine that there is some mystery about it which they can never probe that fear of decimals persists, even though everyone has become accustomed to decimal currency.

It is quite simple to think of numbers written down in columns, even though we do not often do so. A number (222) could be written like this.

Hundreds	Tens	Units
2	2	2

Reading from the right towards the left we have 2 units, or ones (making 2), 2 tens (twenty) and 2 hundreds. Thus the number is two hundred and twenty two. If we put a decimal point on the right of the units column the next column (on the right of the decimal point) represents tenths:

Hundreds	Tens	Units	Decimal point	Tenths
2	2	2		2

In this case the number has become two hundred and twenty two and two tenths but it will be expressed as 222 *point* 2. If we insert another column on the right that will be for hundredths.

Hundreds	Tens	Units		Tenths	Hundredths
2	2	2	·	2	2

Now we have 222 and two tenths and two hundredths, or in other words 22 hundredths. Additional columns could be put on the right to show thousandths, and so on.

It sometimes happens that we insert a zero to the right of the decimal point. If there are no figures on the right of the zero it can be ignored. 1.50 is exactly the same as 1.5 (one and five tenths) because 50 hundredths is the same value as 5 tenths. 1.5 is, in fact, $1\frac{1}{2}$ because there are 2 fives in ten.

Now it can be seen that 0.25 is $^{25}/_{100}$ or $\frac{1}{4}$ and that 0.125 is $^{125}/_{1000}$ or $\frac{1}{8}$.

One of the great advantages of writing fractions as decimals is that they can easily be added or multiplied. We know that 5 plus 4 is 9. In exactly the same way 0.5 plus 0.4 is 0.9

5 + 6 = 11 and if we write this in columns:

Hundreds	Tens	Units
		5
		6
	1	1

it is easy to see that 0.5 + 0.6 = 1.1.

Hundreds	Tens	Units		Tenths
		0	.	5
		0	.	6
		1	.	1

In adding whole numbers it makes no difference if some are in hundreds, some in tens and some less than ten. Anyone can add:

```
125
 25
 50
---
200
```

In the same way, decimal amounts can be added, *provided in each case the decimal point is in the same column*:

$$\begin{array}{l} 0.125 \\ 0.25 \\ 0.50 \\ \hline 0.875 \\ \hline \end{array}$$

It will be seen how much easier it is to add decimal amounts than it would be to add $\frac{1}{8} + \frac{1}{4} + \frac{1}{2}$. The answer is the same because

$$\frac{875}{1000} = \frac{7}{8}$$

The tedious business of reducing fractions to the lowest common denominator is removed because the LCD with decimals is always 10, 100 or 1000.

It is possible, of course, to have numbers containing any number of decimal places, but in everyday life it will very rarely be necessary to deal with any number having more than three figures after the decimal point because the fraction would be so small as to be inconsiderable. Look at a millimetre on a ruler, for example. How often would you need to measure more exactly than that?

If we want to add $\frac{1}{2} + \frac{1}{3} + \frac{1}{5} + \frac{1}{6} + \frac{1}{20}$ we have to work out that the LCD is 60 and then express the sum as:

$$\frac{30}{60} + \frac{20}{60} + \frac{12}{60} + \frac{10}{60} + \frac{3}{60} = \frac{75}{60} = 1\,\frac{15}{60} = 1\tfrac{1}{4}$$

In decimals* the sum would be:

$$0.500 + 0.333 + 0.200 + 0.167 + 0.050 = 1.250$$

The same applies to subtraction

$$\tfrac{3}{4} - \tfrac{1}{5} = \tfrac{15}{20} - \tfrac{4}{20} = \tfrac{11}{20}$$

It is easier to see that $0.75 - 0.20 = 0.55$.

* Once metrication has been completely introduced most of the figures you come across will be whole numbers or decimals but if it is necessary to calculate the decimal equivalent of a vulgar fraction you simply divide the denominator (the lower figure) into 100 or 1,000 or 10,000.

Multiplying is equally easy:

3 × 125	=	375;	3 × 0.125	=	0.375
3 × 5	=	15;	3 × 0.5	=	1.5

You may never have to do multiplication like this and you will not often have to use division (especially if you have a pocket calculator) but if you wanted to compare, for example, two bottles of liquid cleanser, one costing 45p for 18 oz and the other costing 67p for 26⅔ oz (1⅓ pints) it would need quite a complicated calculation to see which was the better buy, and by how much. In fact, the smaller bottle is very slightly cheaper but most people would not bother to work it out.

If the prices were 45p for 500 ml and 67p for 750 ml the calculation would still not be easy (you would have to divide 45 by 5 and 67 by 7.5 to get the price of 100 ml in each case) but it would be easier to understand and less complicated than with imperial measures. However, you may have noticed that the larger bottle was half as big again as the other, so by adding 45p and half of 45p together, you get 67½p, showing that the larger bottle is a little cheaper. Once you have grasped the idea such calculations can be quite fascinating.

To multiply by ten is very easy. If we multiply 4 by 10 the figure moves one place to the left:

	Tens	Units	
		4	× 10
=	4	0	

When we multiply decimals the same thing happens:

	Units		Tenths	
	0	.	4	× 10
=	4	.	0	

So we see that 0.4 multiplied by 10 = 4.0 just as 4 × 10 = 40 or 40 × 10 = 400, 0.04 × 10 = 0.4 and 0.004 × 10 = 0.04.

Some people find it easier, instead of thinking of the *figures* moving to the *left* to think of the *decimal point* moving to the *right*. The effect is exactly the same.

$$0.40 \times 10 = 4.0$$

To multiply by 100 we move the figures *two* places to the left (or move the decimal point two places to the right):

$$0.4 \times 100 = 40.$$

For simplicity up to now we have shown amounts with only whole numbers or decimals but it makes no difference if a number is partly whole numbers and partly decimals.

As an example of addition:

$$2.3 + 4.2 = 6.5$$

In multiplication:

$$5.20 \times 10 = 52.0$$

We saw that a zero (or more than one) on the right of the figures after the decimal point makes no difference to the value.

5.20 is exactly the same as 5.2

But if there is a zero followed by other figures it shows what column those figures belong to: 0.001 means $\frac{1}{1000}$.

Units		Tenths	Hundredths	Thousandths
0	·	0	0	1

0.01 would mean $\frac{1}{100}$.

Units		Tenths	Hundredths
0	·	0	1

It would not make any difference if a zero was put in *afterwards* (ie on the right); 0.010 is exactly the same as 0.01.

Now it is easy to understand how it is possible to express the same amount, even if it is not a whole number of kilogrammes, say, either as kilogrammes or as grammes. 1,500 grammes could be expressed as 1.500 (or 1.5) kilogrammes. In the same way 4.2 litres is the same as 4,200 millilitres.

Since there are 100 centimetres in 1 metre we can similarly express a length either in centimetres or in metres:

250 centimetres = 2.5 (or 2.50) metres.

It works the other way, too. Since there are 10 millimetres in 1 centimetre (because there are 1,000 millimetres in a metre and only 100 centimetres) we can say that:

250 centimetres = 2,500 millimetres, and, using our rule about moving the decimal point:

2,500 millimetres = 2.500 metres.

But since the zeros on the right make no difference 2,500 millimetres = 2.5 metres or 2.50 or 2.500 metres.

One thing it is important to remember is that if we want to add two or more amounts together they must all be converted to the same unit. Suppose we want to add 15 millimetres, 25 centimetres and 2 metres we can say either:

(a) 15 millimetres = 1.5 centimetres and 2 metres = 200 centimetres, so the addition becomes:

1.5	centimetres
25.0	centimetres
200.0	centimetres
226.5	centimetres

or (b) 25 centimetres = 250 millimetres,
2 metres = 2000 millimetres.
Now the sum becomes

15	millimetres
250	millimetres
2,000	millimetres
2,265	millimetres

or (c) 15 millimetres = 0.015 metres

25 centimetres = 0.25 metres.
In this case the addition becomes:

0.015	metres
0.25	metres
2.0	metres
2.265	metres

We know from the rules about multiplying by 10 that 2.265 metres = 226.5 centimetres = 2,265 millimetres so that it makes no difference what unit of measurement we add up in *provided each figure is converted to that unit.*

To take another example, suppose we had four packets of butter weighing 1½ kg, 1 kg, 200 g and 40 g respectively, what would be the total weight?

(a)

1½ kg	=	1.5 kg
1 kg	=	1.0 kg
200 g	=	0.2 kg
40 g	=	0.04 kg
Total		2.74 kg

or (b)

1½ kg	=	1,500 g
1 kg	=	1,000 g
200 g	=	200 g
40 g	=	40 g
Total		2,740 g

2,740 g = 2.740 kg (or 2.74 kg)

To sum up

1. When adding figures containing decimals you can ignore any zeros on the right of the decimal point if there are no more figures afterwards.

2. You must take into account a zero in the decimal number if there are figures on the right of it.
3. If the figures are written down so that all the decimal points come under each other the addition can be done in the ordinary way with the decimal point in the answer exactly underneath the other decimal points.
4. If you are adding amounts expressed in different units you must convert them all to the same unit before adding them up.

13. How Much Do You Remember?

(If the exact answer is not shown choose the nearest of the alternatives.)

1. How many inches are there in a metre?
   (a) 33 (b) 36 (c) 39 (d) 42

2. How many centimetres are there in a metre?
   (a) 50 (b) 100 (c) 200 (d) 1,000

3. How many centimetres are equivalent to 4 inches?
   (a) $2\frac{1}{2}$ (b) 5 (c) $7\frac{1}{2}$ (d) 10

4. What is the metric equivalent of 440 yards?
   (a) 350 m (b) 400 m (c) 48 m (d) 500 m

5. What is the metric equivalent of 30 mph?
   (a) 25 km/h (b) 40 km/h (c) 50 km/h (d) 60 km/h

6. How many millilitres are there in a litre?
   (a) 100 (b) 500 (c) 1000 (d) 10,000

7. How many pints are there in a litre?
   (a) $\frac{1}{2}$ (b) $1\frac{1}{4}$ (c) $1\frac{1}{2}$ (d) $1\frac{3}{4}$

8. How many litres are there in a gallon?
   (a) $4\frac{1}{2}$ (b) 5 (c) 10 (d) 20

9. What is nearly equal to one ounce?
   (a) 5 g (b) 10 g (c) 28 g (d) 56 g

10. What is equivalent to 15 kg?
    (a) 10 lb (b) 20 lb (c) 25 lb (d) 33 lb

11. What is equivalent to 1 cwt?
(a) 25 kg (b) 50 kg (c) 75 kg (d) 100 kg

12. What is the height of a normal door?
(a) 1½ metres (b) 2 metres (c) 3 metres (d) 4 metres

13. What is the Celsius equivalent of 32°F?
(a) – 243°C (b) – 32°C (c) 0°C (d) 35°C

14. What is the normal body temperature?
(a) 37°C (b) 45°C (c) 98°C (d) 100°C

15. If 9 apples weigh 2 lb what would you expect 20 apples to weigh?
(a) 500 g (b) 1 kg (c) 2 kg (d) 5 kg

16. How many grammes are there in a kilogramme?
(a) 10 (b) 100 (c) 1,000 (d) 10,000

17. How many decilitres are there in a litre?
(a) 10 (b) 100 (c) 1,000 (d) 10,000

18. How many millimetres are there in a metre?
(a) 10 (b) 100 (c) 1,000 (d) 10,000

19. What is the weight of 1cc of water?
(a) 1 gramme (b) 10 grammes (c) 100 grammes (d) 1 kg

20. How many kilogrammes are there in a tonne?
(a) 50 (b) 100 (c) 500 (d) 1000

Answers to Quiz

1 c; 2 b; 3 d; 4 b; 5 c; 6 c; 7 d; 8 a; 9 c; 10 d; 11 b; 12 b; 13 c; 14 a; 15 c; 16 c; 17 a; 18 c; 19 a; 20 d.

14. Tables

Square and cubic measures

Table 1A. Imperial to metric

Imperial	Metric
1 sq inch	645 mm^2 (sq millimetres)
1 sq foot	9290 mm^2 (sq millimetres)
1 sq yard	0.836 m^2 (sq metres)
1 acre	0.405 ha (hectares)
1 cu inch	16.387 cm^3 (cubic centimetres)
1 cu foot	0.0283 m^3 (cubic metres)
1 cu yard	0.765 m^3 (cubic metres)

Table 1B. Metric to imperial

Metric	Imperial
1 mm^2 (sq millimetre)	0.0016 sq inch
1 cm^2 (sq centimetre)	0.1566 sq inch
1 m^2 (sq metre)	1.196 sq yards
1 ha (hectare)	2.471 acres
1 cm^3 (cubic centimetre)	0.061 cu inch
1 m^3 (cubic metre)	1.308 cu yards

For capacity see Tables 2A and 2B.

Table 2A. Imperial to metric

Imperial		Metric	
	1 fl oz	28.41 ml	(2.8 cl)
	2 fl oz	56.83 ml	(5.7 cl)
	3 fl oz	85.24 ml	(8.5 cl)
	4 fl oz	113.66 ml	(1.14 dl)
(1 gill)	5 fl oz	142.07 ml	(1.42 dl)
(½ pt)	10 fl oz	284.13 ml	(2.84 dl)
(1 pt)	20 fl oz	568.26 ml	(5.68 dl)
(2 pt)	1 quart	1136.52 ml	(1.137 litre)
(4 pt)	(½ gal)	2.273	litre
(8 pt)	(1 gal)	4.546	litre
	(2 gal)	9.092	litre
	(4 gal)	18.184	litre
	(5 gal)	22.730	litre
	(10 gal)	45.461	litre
	(50 gal)	227.305	litre
	(100 gal)	454.610	litre

To those with a keen eye for detail it may seem puzzling that although 1 litre of water weighs 1 kilogramme – and therefore 1 millilitre weighs 1 gramme – 1 ounce in *weight* is equivalent to 28.35 grammes whereas 1 *fluid* ounce of water is equal to 28.41 millilitres, which would weigh 28.41 grammes, a difference of about 0.2 per cent.

The reason is that 1 ml of water weighs 1 g at 4 °C while 1 fluid ounce weighs 1 oz avoirdupois at 62 °F (16.66 °C). 1 oz (in weight) of water would occupy a smaller space than the same weight of water at 16.66 °C. Or it could be said that 1 ml of water at 16.66 °C would be less dense, and therefore would weigh less, than 1 ml of water at 4 °C. This is the temperature at which water is at its maximum density; water expands as

Table 2B. Metric to imperial

Metric		Imperial
1 ml		0.035 flo oz
2 ml		0.07 fl oz
3 ml		0.11 fl oz
4 ml		0.14 fl oz
5 ml		0.18 fl oz
10 ml	(1 cl)	0.35 fl oz
15 ml		0.53 fl oz
20 ml		0.7 fl oz
25 ml		0.88 fl oz
30 ml		1.06 fl oz
40 ml		1.41 fl oz
50 ml		1.76 fl oz
100 ml	(1 dl)	3.52 fl oz
500 ml		17.61 fl oz
600 ml		1 pint 1.1 fl oz
1000 ml	(1 litre)	1 pint 15.2 fl oz
	(2 litres)	3 pints 10.4 fl oz
	(4 litres)	7 pints 0.8 fl oz
	(5 litres)	1 gal 16.1 fl oz
	(10 litres)	2 gals 1 pt 12.2 fl oz
	(50 litres)	11 gals 1.1 fl oz
	(100 litres)	22 gals 2.2 fl oz

the temperature rises but it also expands as it freezes. This sometimes causes pipes to burst. The fracture is not usually detected until the ice melts but the damage was done by the expansion of the water as it became ice, not by the thaw.

Linear measure

Table 3A Imperial to metric

Imperial	Metric	Approximate
1 inch	2.54 cm	2½ cm
2 inches	5.08 cm	5 cm
4 inches	10.16 cm	10 cm
6 inches	15.24 cm	15 cm
8 inches	20.32 cm	20 cm
10 inches	25.40 cm	25 cm
12 inches (1 foot)	30.48 cm	30½ cm
2 feet	0.61 m	61 cm
3 feet (1 yard)	0.91 m	9/10 m
5 feet	1.52 m	1½ m
10 feet	3.05 m	3 m
10 yards	9.14 m	9 m
20 yards	18.29 m	18¼ m
50 yards	45.72 m	45¾ m
100 yards	91.44 m	91½ m
200 yards	182.88 m	183 m
220 yards	201.16 m	201 m
440 yards (¼ mile)	402.32 m	402 m
500 yards	457.20 m	457 m
880 yards (½ mile)	804.67 m	805 m
1,000 yards	914.40 m	914 m
1,760 yards (1 mile)	1,609.34 m	16 km

The approximate figures should not be added together or multiplied.
The figure 0.3048 m = 1 foot is an exact conversion.

Table 3B. Metric to imperial

Metric	Imperial	Approximate
1 mm	0.0396 inch	40 thou
1 cm	0.3937 inch	0.4 inch
2 cm	0.7874 inch	0.8 inch
5 cm	1.9685 inch	2 inches
10 cm	3.937 inches	4 inches
20 cm	7.874 inches	8 inches
30 cm	11.811 inches	1 foot
1 m	1.094 yards	$1\frac{1}{10}$ yards
2 m	2.187 yards	$2\frac{1}{5}$ yards
5 m	5.468 yards	$5\frac{1}{2}$ yards
10 m	10.936 yards	11 yards
20 m	21.872 yards	22 yards
50 m	54.680 yards	55 yards
100 m	109.361 yards	110 yards
200 m	218.722 yards	220 yards
400 m	437.445 yards	$\frac{1}{4}$ mile
500 m	546.807 yards	550 yards
1,000 m (1 km)	1,093.613 yards	1,100 yards
5,000 m (5 km)	5,468.065 yards	3 miles
10,000 m (10 km)	10,936.130 yards	6 miles

The approximate figures should not be added together or multiplied.

Nautical miles, cable-lengths and fathoms

Traditionalists, especially those with a nautical background, may object to the phasing out of the nautical mile and its derivatives, the cable-length and the fathom, because of their relationship to the earth's circumference.

At the equator the circumference of the earth is 24,901.8 English miles (40,067 km), divided into 360 degrees of longitude, each of 69.17 English miles, equivalent to 60 geographical miles. Since there are 60 minutes in a degree, one geographical mile (6,087 feet) is equal to 1 minute of longitude. One traditional English nautical mile is equal to 6,082.66 feet but the international nautical mile (official unit in the US since 1954) is 6,076.1033 feet, or 1.852 km.

One cable-length is one-tenth of a nautical mile (ie 608 feet) and 100 fathoms make 1 cable-length, but in practice the fathom (originally the reach of a man's outstretched arms) is 6 feet (1.83 m), not 6.08 feet (1.85 m) which one-thousandth of a geographical mile would be.

Measures of weight

Table 4A. Imperial to metric

Imperial	Metric	Approximate
1 oz	28.35 grammes	30 g
4 oz	113.40 grammes	115 g
8 oz	226.80 grammes	225 g
1 lb	453.59 grammes*	450 g
2 lb	907.18 grammes	900 g
5 lb	2.268 kilogrammes	2¼ kg
10 lb	4.53 kilogrammes	4½ kg
14 lb (1 stone)	6.34 kilogrammes	6 kg
20 lb	9.07 kilogrammes	9 kg
50 lb	22.68 kilogrammes	23 kg
56 lb (½ cwt)	25.40 kg	25 kg
112 lb (1 cwt)	50.80 kg	50 kg
2,240 lb (1 ton)	1,016.04 kilogrammes	1,000 kg (1 tonne)

The approximate figures should not be added together or multiplied.
* The exact factor is 453.59237

The uncertainty about the exact circumference of the earth – and hence the length of one degree of longitude – is due to the fact that the world is not a perfect sphere.

All this will be of little consequence except to astronomers and navigators and in the electronic age they have more sophisticated means of determining distances on the earth's surface and more accurate ways of measuring the depth of the sea than the lead-line.

Table 4B. Metric to imperial

Metric	Imperial	Approximate
1 g (gramme)	0.035 oz	1/30 oz
100 g (grammes)	3.53 oz	3½ oz
500 g (grammes)	17.64 oz	1 lb
1 kg (kilogramme)	2.205 lb	2¼ lb
2 kg (kilogrammes)	4.410 lb	4½ lb
5 kg (kilogrammes)	11.025 lb	11 lb
10 kg (kilogrammes)	22.051 lb	22 lb
20 kg (kilogrammes)	44.101 lb	44 lb
50 kg (kilogrammes)	110.25 lb	1 cwt
100 kg (kilogrammes)	220.50 lb	2 cwt
200 kg (kilogrammes)	441.01 lb	440 lb
500 kg (kilogrammes)	1,102.3 lb	½ ton
1,000 kg (1 tonne)	2,204.6 lb	1 ton
The approximate figures should not be added together or multiplied.		

Calculate your own height on the scale shown here and then, as a check, verify it with a metric rule or tape measure. All you have to remember is that 100 centimetres is equal to one metre.

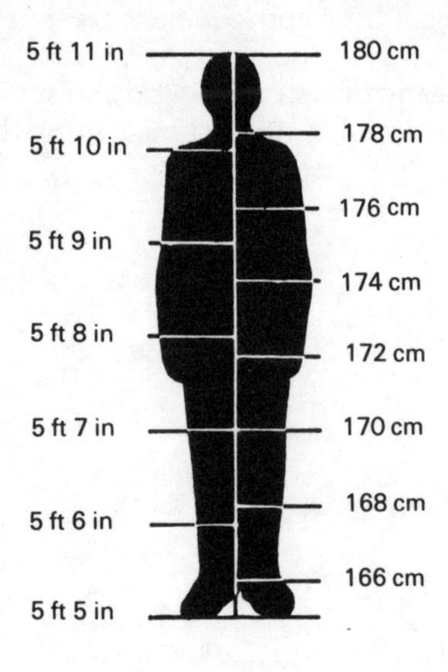